Deserts in Europe:

None

Deserts in Asia:

3. Gobi
4. Arabian
5. Great Sand
6. Great Salt
7. Syrian
8. Thar
9. Turkestan
10. Takla Makan

Deserts in Africa:

16. Sahara
17. Somali
18. Kalahari
19. Namib

Deserts in Australasia:

11. Gibson
12. Great Sandy
13. Great Victoria
14. Simpson
15. Tanami

Facts about sand dunes and sand seas

An eighth of the world's land surface is desert, large areas of which are covered with sand dunes. The world's largest desert is the Sahara of North Africa which covers over 3 million square miles. Sand dunes cover about a quarter of this desert, the rest is bare rock or plains covered in pebbles.

Deserts with sand dunes are found in many continents, either because the desert is a long way from the sea (as in the Gobi desert), or because the desert is sheltered from rain-bearing clouds by high mountains (as in Patagonia, Argentina). Most are found in a broad belt around the world centered on 25 degrees north and south of the equator.

Sand seas covernearly one million square miles in the Sahara desert. Some sand dunes rise over 1000 ft above the surrounding plains. The area with the highest sand dunes is in Algeria and is part of the sand sea called the Grand Erg Occidental.

The largest area in the world directly scoured by blowing sand lies near the Tibesti mountains in the Sahara desert. It covers 35,000 square miles. Huge sand seas lie just downwind of this scoured area.

Grolier Educational Corporation

SHERMAN TURNPIKE, DANBURY, CONNECTICUT 06816

LAND SHAPES

DUNE

Author
Brian Knapp, BSc, PhD
Art Director
Duncan McCrae, BSc
Editor
Rita Owen
Special models
*Tim Fulford, MA, Head of Design and Technology,
Leighton Park School*
Illustrator
David Hardy
Print consultants
Landmark Production Consultants Ltd
Printed and bound in Hong Kong
Produced by
EARTHSCAPE EDITIONS

First published in the USA in 1993 by
GROLIER EDUCATIONAL CORPORATION,
Sherman Turnpike, Danbury, CT 06816

Library of Congress #92–072045

Cataloging information may be obtained
directly from Grolier Educational Corporation

Title ISBN   0–7172–7182–X

Set ISBN   0–7172–7176–5

Acknowledgements. The publishers would like to
thank the following: Leighton Park School, Martin
Morris and Redlands County Primary School.

Picture credits. All photographs from the
Earthscape Editions photographic library except
the following (t=top, b=bottom, l=left, r=right):
David Higgs 25; NASA 27t; ZEFA 8/9, 16/17,
24/25, 33b, 34b, 35t.
Cover picture: Death Valley, California, USA.

In this book you will find some
words that have been shown in **bold**
type. There is a full explanation of
each of these words on page 36.

On many pages you will
find experiments that you
might like to try for
yourself. They have been
put in a blue box like this.

In this book mi means miles and
ft means feet.

These people appear on a number
of pages to help you to know the
size of some landshapes.

CONTENTS

World Map   2

Facts about sand dunes   4

Introduction   8

Chapter 1: Where sand comes from

Patterns of sand   10

Where desert sand comes from   12

The importance of sandstorms   14

Where sand settles   16

Chapter 2: Types of sand dune

Crescent-shaped dunes   18

How sand dunes move   20

Why dunes have steep sides   22

Barchans   24

Sand ridges   26

Sand mountains   28

Sand seas   30

Chapter 3: Sand deserts of the world

The Sahara Desert   32

Sorting sand grains   34

New words   36

Index   37

Introduction

Sand dunes are mounds of sand that have been shaped by blowing wind. They are most common in deserts – lands with little water – because there are few plants to protect the soil from strong winds.

During a strong wind **dust** can form into great clouds that reach high into the sky. The heavier particles, like sand, do not blow away like dust, but rather they hop and bounce along near the ground, where they begin to form into dunes.

People who have visited sand deserts often say that sand dunes look rather like waves that have been 'frozen' in a sea. This is why large areas of sand dunes are known as **sand seas**.

Sand dunes come in many shapes and sizes, each shaped and reshaped by the wind. As there is nothing to hold the sand **grains** in place, dunes rarely remain the same from one week to the next.

Sand dunes are mostly formed in the harshest places on Earth. They are exciting to look at and easy to understand. In this book you can find out where the sand comes from, how it moves and the many shapes it forms into. Enjoy exploring the landshapes of dunes by turning to any page you choose.

Dune buggies are vehicles that have been designed specially for travel over sand. Can you see how they can cope with loose sand?

Take care in the sand dunes

Sand dunes are some of the world's most exciting landshapes and you are sure to want to visit them. But never go into an area of sand dunes without taking a supply of water, clothes to protect your head from the glaring sun and thick shoes to keep the heat from your feet. Above all, only go in the company of an adult. Sand dunes in deserts can be dangerous places for the unwary and deaths have occurred because people have gone walking unprepared.

Chapter 1

Where sand dunes come from

Patterns of sand

Each dune began as rocks in far distant mountains. These rocks were broken down, or **weathered**, into grains and then carried by floods and scattered on the desert **plains**. From here the wind took over, driving the sand on until it came to rest in a sand sea.

Each of these larger shapes is called a sand dune (see page 18).

The steep, or fall, side of a dune gives clues to the way sand has settled (see page 22).

The gentle, or ramp, side of a sand dune shows which way the wind blew (see page 20).

These layered rocks clearly show the effects of heavy rain, with deep channels being cut from top to bottom of the cliff.

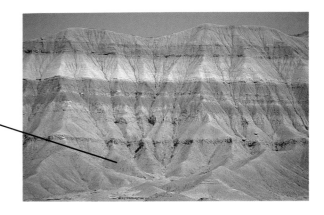

Sand dunes need rocky mountains to keep them supplied with sand (see page 12).

Sand, gravel, stones and dust are often washed down the valleys and out onto plains during **flash floods**. Strong winds sort the sand and dust and leave the pebbles behind (see page 14).

The small ridges of sand found on the surfaces of dunes are called ripples (see page 26).

Sometimes sand dunes grow into sand mountains (see page 28).

Where desert sand comes from

Sand is the name for a particular size of rock fragment. Desert sand begins as rocks on bare mountain slopes that are affected by extreme temperatures. Each day they are baked and each night they are cooled. From time to time they are also drenched in rainwater from torrential storms. The effect of this is to break up the rock into smaller and smaller particles and eventually to sand.

This deep valley, or **canyon**, gathers water and rock waste which is flushed to the nearby plains.

Cascades that carry sand
From time to time, torrential thunderstorms break over the mountains. Water rushes down from the mountains into the valleys carrying millions of tons of broken rock, stones, sand and mud and spreads them over the flat plains beyond. Later, when the rock debris has dried out, the wind can pick up the finer surface particles such as dust and sand, and carry them away.

These are the slopes that provide the sand.

Break up

The source of sand is often a bare rock face like the one shown in the background of this picture. Heating by day and cooling by night make the rock swell and shrink and this helps to weaken it. When it rains, water soaks into the surface of the rock and this may dissolve the **cement** that holds the grains together.

Rock flakes and grains that have been loosened from the solid rock can easily be washed down to the plains during a storm.

This is what desert sand grains look like. They are less than $\frac{1}{10}$ th inch across but bigger than dust.

Here are the thick spreads of rock debris that have been left behind when the flood has subsided.

This is what the debris from a flash flood looks like. Fierce sandstorms are needed to whisk the sand from among the stones.

SCALE

The importance of sandstorms

The first step to making sand dunes occurs when the weather breaks mountain rocks into smaller pieces and then scatters them over desert plains. But the broken debris is a mixture of coarse and fine grains and these must next be sorted by the wind.

Sandstorms – when strong winds whisk millions of tons of sand and dust into the air – are the most powerful means of sorting sand from other sizes of material in a desert. But sand carried in the wind also has great power to scour the rocks it is flung against, as many landshapes show.

Wind has to be very strong to whisk grains from among the other materials on the desert floor.

Watch sand move
Fill a shallow tray with sand and place a sheet of white paper behind it.

The idea is to see how high sand can hop when it moves.

Now use a fan to blow gently over the top of the sand. Nothing moves. Try raising the wind speed until the sand does move. You should be able to see the sand skipping over the surface, making long hopping movements as it is pulled from the surface, then let to fall back again.

14

These cans have been sand-blasted during a single sandstorm. It shows just how abrasive moving sand can be.

Sand-blasting

Sand grains can be blown fiercely against rock during a sandstorm. This in turn may release other sand grains. The effects of wind **erosion** can be seen in the tell-tale signs like the ones shown on this rock below.

This piece of rock is more than 6 ft above the ground and it is too high for the wind to carry sand. As a result it has kept its rugged shape.

SCALE

Sand grains carried by the wind have scoured the base of this rock making it smooth. This process, called **abrasion**, has produced even more fine material to build distant sand dunes.

Where sand settles

Sand builds up in places where the wind can no longer carry it. An area of sand dunes, no matter how big, is really a gigantic sand pit, a place from which sand cannot escape.

You can see why sand should build up in some places and not others when you walk along a windy city street. You see places where the leaves and paper rubbish are being blown along, and then, from time to time, you see places where the wind slackens and the dirt settles out. The landscape has its windy places on the mountains; its sand pits lie in places where the winds are less fierce.

This picture will help you to appreciate the way that sand dunes rest on rocky desert plains. You can also see how tall sand dunes can be. This village is built right against enormously high sand dunes. Would you like to live so close to a moving landshape like this?

Because there is so much sand in these dunes, the wind drives them forward very slowly. In many deserts the wind blows one way for part of the year, and then blows from the opposite direction for another part of the year. As a result there may be no overall movement at all.

Chapter 2

Types of sand dune

Crescent-shaped dunes

There are various patterns of sand dune found in deserts. The most common type is shown on this page and it is of medium size, each crescent being perhaps a hundred yards long. When many crescent-shaped dunes occur together they form a **chaotic** pattern.

Crescent-shaped dunes are produced when the wind blows in many different directions during the year.

The long wavy ridge top, or crest, of the dune shows clearly in this picture. Notice how the steep faces of the dunes also change from one side to the other. This is the result of many changes in wind direction.

This is the fall side of the dune. It is a long, straight slope and there are no ripples on it.

Features of sand dunes

Most sand dunes form part of a large pattern of sand. Most dunes have crescent-shaped tops that change direction from one dune to the next. The sharp, curved ridge is the dune crest.

These dunes can reach tens of yards in height and each crescent may be up to a hundred yards long.

Each has a more gentle side, called a ramp, and a steeper, or fall, side.

The surface of most sand dunes is covered in ripples.

This is the ramp side of the dune. It is much more gentle than the fall side and it is covered with ripples.

Make sand dunes

Find out about how sand moves and the patterns it makes by using very dry sand and a hair dryer. The picture above shows a simple pattern made with a hair dryer. You may care to experiment to see if you can imitate the sand dune patterns shown in the picture on the left and on later pages.

How sand dunes move

No matter from which direction the wind blows, and no matter how strong it is, dunes are never blown away. Even the strongest winds do not flatten dunes because dunes are *made* by the wind.

However, dunes are always changing shape and moving as the wind blows sand from one side of the crest to the other. The sand is blown up the gentle, or ramp side, across the crest and then it simply spills onto the fall face. As a result no sand is lost from the dune, it simply moves forward a little. The stronger the wind blows the faster the dune moves.

The sand on a ramp is firm because it is piled at a low angle. Notice how these foot prints are only shallow – compare them with the foot prints on page 23.

The footprints are on the ramp side of the dune where the sand is firm. The fall side is on the left.

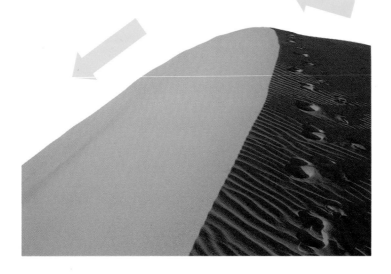

Sometimes plants can stop the movement of dunes, but moving dunes can bury even trees – as this picture shows.

Watch sand move across a dune

To see how the sand moves fill a tray with sand and make a dune ridge across it. The size and shape of the dune is not important, but the sand must be completely dry. Fine sand works best.

Use a fan to produce wind strong enough to make the sand move, then watch carefully to see the dune change shape.

After a while, try moving the fan so that the wind blows from a slightly different direction. Does the dune shape change too?

Fan to give a steady wind.

Tray with completely dry *fine* sand.

Ridge made into a 'dune' about 3 inches high.

Why dunes have steep sides

Sand dunes have a steep side as well as a gentle side. This is where the sand comes to rest after it has been blown up the ramp in a sandstorm. Even though the slope is steep, it will not become gentle. Even if it is trampled down, it will gradually build back again during the next sandstorm. This is why it happens.

People can enjoy the steep slope of a sand dune for riding sand buggies (see page 9) or for playing games. This will not harm the landshape because it is rebuilt in the next sandstorm.

Secret to sand piles

Sand is made of many small grains which are shaped roughly like tiny volley balls.

Sand grains therefore do not fit together very well and they tend to roll down a slope. When a slope is steep they will bounce down it; when the slope is more gentle they will more easily come to rest.

Find out how sand falls

Fill a tray with clean dry sand. The sand must be dry enough to run through your fingers. Now put a glass jar on the tray and fill it with sand. Pull the jar up gently, watching the way the sand falls out.

You could measure the angle of the sand slope by using a protractor. Try repeating the experiment. Does the sand always fall at the same angle?

This picture shows the way the sand falls down as a person walks up the steep face of a sand dune. Compare how easily the sand moves with the picture on page 20.

Barchans

These are a very special type of sand dune that form in deserts when the winds are just right.

Instead of being in long rows, barchans are separate dunes that make a fascinating pattern.

How barchans form

Barchans always occur in groups and are referred to as a field of barchans. Each barchan is of a similar size and is fairly evenly spaced apart. This tells us that the barchans are formed by some special pattern of winds.

A barchan has a fat 'body' and a pair of 'horns'. The horns show which way the winds normally blow, because the tips of the horns point downwind. However, from time to time strong winds blow at right angles to the normal direction. These winds make sure that the dunes do not meet up and form long ridges.

Give a barchan horns

A barchan is a sand dune with 'horns'. You can see how a barchan moves by making a short ridge of sand on a tray of dry sand.

Use a fan to make a wind that will blow sand along. Soon you will see the end of the ridge begin to move forward. These are the places with the smallest amounts of sand to move and they quickly start to form the horns that you get on a barchan.

wind direction

Sand ridges

When sand moves it often makes patterns. The smallest ridges and troughs are only a few inches across. They are like mini-dunes and they are called ripples.
The largest ones can be tens of miles long and a hundred feet high. They are called seif dunes.

Whatever their size, long ridges of sand are caused by strong swirling movements of wind that pile sand up into straight crests.

Look at the ripples in this picture. Can you see how there appears to be a pattern, with ripples spaced almost evenly across the back of the sand dune? The ripples are formed during a sandstorm when sand hops over the surface.
 Ripples do not form when the winds are light. Between sandstorms some grains will be blown from the crests into the troughs and the ripples will begin to fade away.

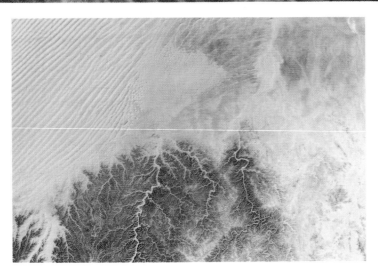

Seif dunes

Seif dunes are long ridges of sand that are found in places where the wind always blows in the same direction. The ridges form in the same direction as the wind. This picture was taken from space – the view shows almost a hundred miles of seif dunes in the Saudi Arabian desert! You can see a dried up river course in the bottom of the picture.

This person is standing on a field of ripples. Ripples are always just a few inches across, but their pattern changes depending on the strength and direction of the wind.

Sand mountains

Sometimes the wind blows sand into huge piles called sand mountains. Some of these piles of sand may be over 1000 ft high and tower above the other dunes of a sand sea.

How sand mountains form

Sand mountains form in places where the wind often changes direction. It tends to sweep sand dunes across each other and sometimes they pile up and make a mountain, just as sometimes sea waves cross each other and form an extra-large wave.

Once a sand mountain has grown, it changes the pattern of winds in such a way that more sand mountains can grow nearby. In this way sand mountains often form in patterns.

28

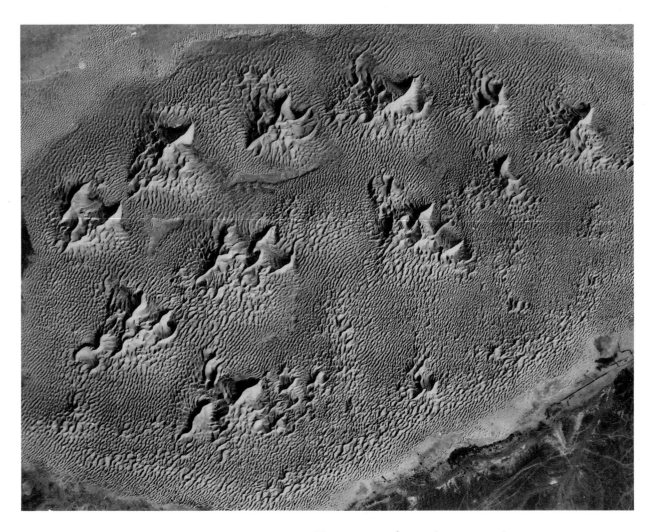

Patterns of sand mountains

The picture above shows some sand mountains from the air. The sand mountains are the star-like shapes, looking almost like icing decorations on a cake.

As you can see, the mountains form a pattern, with each mountain surrounded by a sea of sand dunes.

Unlike sand dunes, sand mountains have steep slopes on every side. This picture shows the ridge leading to the mountain summit.

Sand seas

In a sand sea you can find most of the types of sand dunes that have been described on the earlier pages. Some sand seas are very large and stretch for hundreds of miles (see page 32). Over such large areas the wind patterns change, so that a pattern of dunes in one part of a sand sea may not be at all like the sand dunes elsewhere.

From the top of a sand mountain the sand sea made by the dunes shows clearly. Ripples, dunes and sand mountains are all made and moved by the wind, just as sea waves and breakers form and reform in an ocean.

Imitating sand sea forms

Because it takes so long for sand to move, it is often easier to imitate some of the sand patterns by using a tank of water.

You need two or three people to act as sources of wind for this activity.

When only one person blows the water across the tank, the pattern of waves is very simple. Notice that the waves grow higher as the wind speed increases.

See what happens when more people blow air from other directions. With three people blowing you should even be able to make water mountains!

Chapter 3
Sand deserts of the world

The Sahara Desert

The northern part of Africa is one of the driest regions on Earth. It is called the Sahara Desert. It covers three million square miles or about the size of Western Europe.

Although much of the Sahara consists of barren plains of rock and pebbles, or high rugged mountains, a quarter of it is made of sand dunes. Vast sand seas cover almost a million square miles.

The road in the picture on the right gives you some idea of the immense size of the Sahara. The picture has been taken from the edge of a rocky **plateau**, like the ones seen in the distance. Notice how much sand lies everywhere. It has only been shaped into dunes near the left-hand edge of the picture.

Date palm trees are a sign that there is moisture below the surface – even though they may be surrounded by sand dunes. A source of water in a desert is called an oasis.

Finding sand seas in the Sahara

Sand seas are common in the Sahara and each one has a special name. On atlases you will find them called *erg*, from the Arabic word for sand sea. You may care to trace a map of north Africa from an atlas and write down the number of names that have erg in them. This will tell you where the sand seas are to be found.

Part of the Grand Erg Occidental.

Sorting sand grains

The Sahara has every possible type of sand dune. In the picture below you can see some low crescent-shaped dunes in a valley and sand mountains behind. Many types of dune are made from different sizes of sand grains.

Sand mountains, probably made of coarser sand.

This dwindling river and lake are all that remain of a recent flash flood.

Managing landshapes

People who live in the Sahara sometimes try to stop the sand from moving. To do this they plant vegetation specially adapted to grow in dry places. In the picture on the right the vegetation has been made into a kind of wind-break to protect a section of road. By looking for the road, can you say how successful this has been?

Low, crescent-shaped dunes in the valley bottom, probably made of finer sand.

Most people regard the sandy deserts as impassable because of the way sand moves about. Just think what an effort it would be for a road-making bulldozer to make an impression on the massive dunes shown in the picture below.

New words

abrasion
the wearing away, or scouring of a rock as sand grains are thrown against it during a sand storm

canyon
a gorge-like valley found in dry areas. Canyons have bare, rocky sides with no soil cover. Other names for canyon include arroyo (US) and wadi (north Africa)

cement
the natural material that holds sand grains and other particles together to make a rock. Lime is a common cement and is easily dissolved when it gets wet, allowing the sand grains to be washed or blown away

chaotic
from chaos, a word meaning something that has no pattern or sequence to it

dust
the fine pieces of rock waste that are easily caught by the wind. People sometimes use the words clay and silt to mean dust. When dust becomes wet it turns into mud

erosion
the breaking down and then carrying away of the rocks or soil by a combination of weather, landslides, wind and rivers

flash flood
a sudden storm in a desert may cause large amounts of water to flow from rocky hills and gather on the nearby plains and in river beds. There is no time for the water to sink in to the ground, and instead it floods out across the countryside in a matter of minutes, often without any kind of warning

grains
particles of rock waste, often of a single type of material, that are small enough to be readily moved by wind or water. Sand particles are called grains; bigger particles such as gravel or pebbles, are not considered to be grains

plain
a low-lying area. Plains are not absolutely flat, but have very gentle slopes. Most plains have been formed by long periods of land erosion by rivers

plateau
a name for a tableland, or flat high land

sand-blasted
the process of cleaning metal in a factory using sand grains blown from a high pressure air gun. Nature can produce much the same effect during a sand storm

sand sea
the large areas of sand dunes that are found in the world's great deserts. Sand seas would normally cover many thousands of square miles

sandstorm
a time when the winds blow so strongly that sand grains can be carried along. A duststorm is far more common than a sandstorm, and dust is always lifted into the air when winds are strong enough to move sand. Dust gets in the eyes and mouth, but it does not erode the land; sand scours rocks up to about six feet above the general level of the plains

weathered
a rock is weathered by elements such as frost, sun or rain, causing it to break up or fall apart

Index

abrasion   15, 36
Arabian desert   3
Atacama desert   2

barchan   24

canyon   12, 36
cement   13, 36
chaotic   18, 36
crescent-shaped dunes
        18, 34
crest   18, 20

date palm   32
Death Valley   2
desert   10
dune buggies   9
dunes   8
dust   8, 11, 36

erg   33
erosion   15, 36

fall face   10, 18, 20
flash flood   11, 34, 36
flood   10

Gibson desert   3
Gobi desert   4
grains   22, 36
Great Salt desert   3
Great Sandy desert   3
Great Victoria desert   3

Mojave desert   2

oasis   32

Patagonian desert   2, 4
plain   17, 36
plateau   32, 36

ramp   10, 18, 20
ripples   11, 26, 30

Sahara desert   3, 4, 27,
        32
sand   8
sand buggies   23
sand grains   13, 15
sand mountains   11, 28,
        34
sand sea   8, 30, 36
sand-blasted   15, 36
sandstorm   14, 15, 36
Sechura desert   2
seif dunes   26, 27
Simpson desert   3
Somali desert   3
Sonora desert   2

Takla Makan desert   3
Tanami desert   3
Thar desert   3
Tibesti mountains   4
Turkestan desert   3

weathered   10, 36
wind-break   35